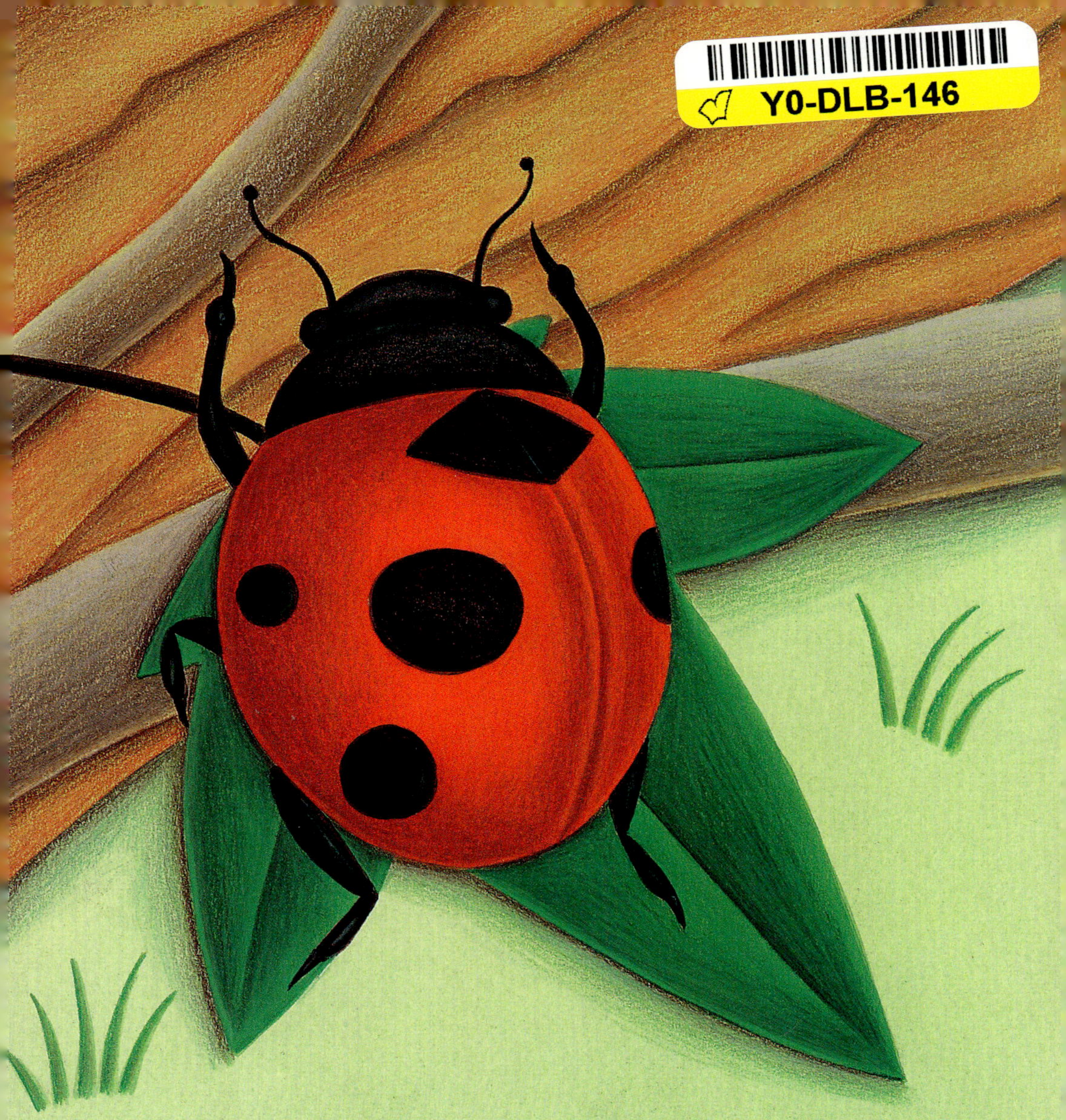

I am a bug.

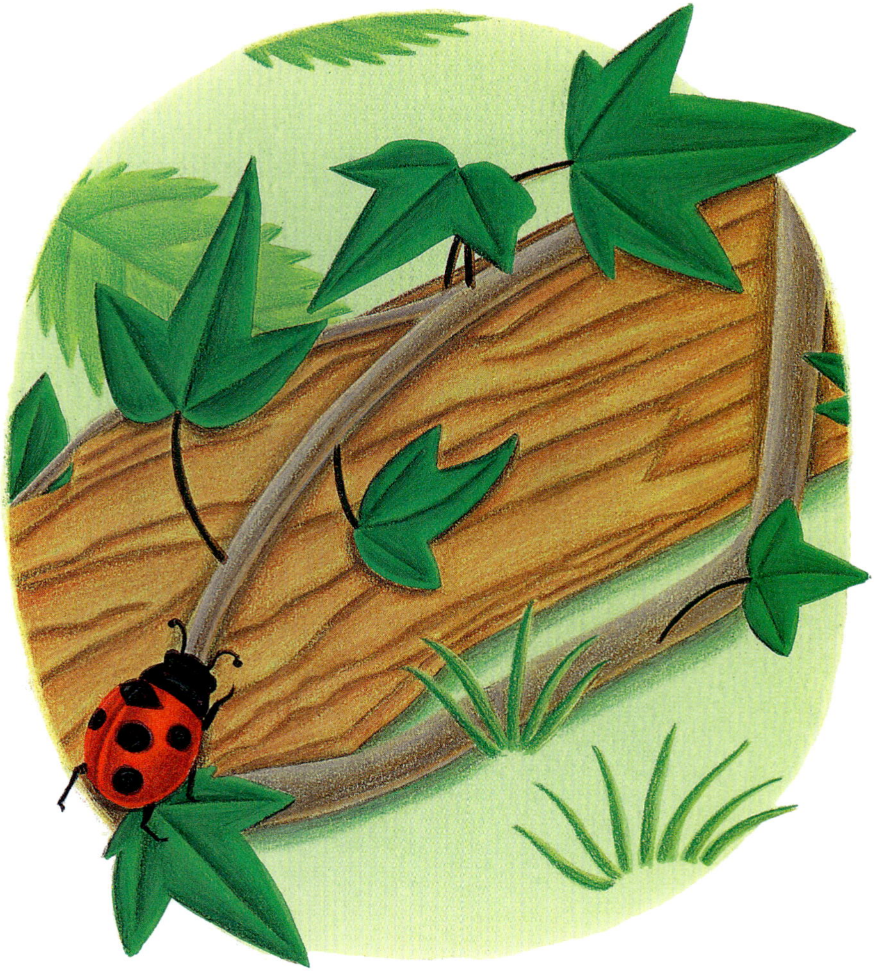

I run up.

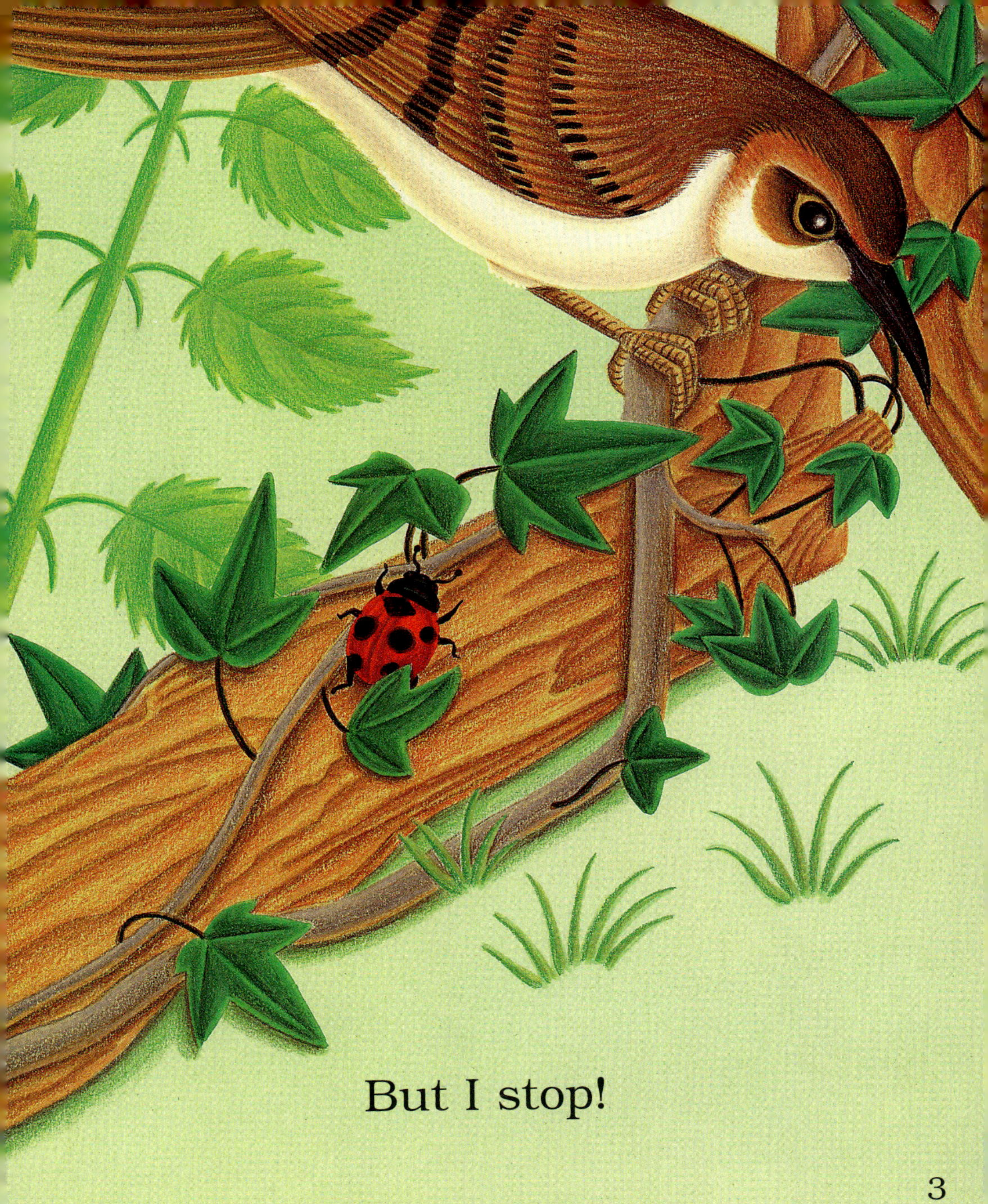

But I stop!

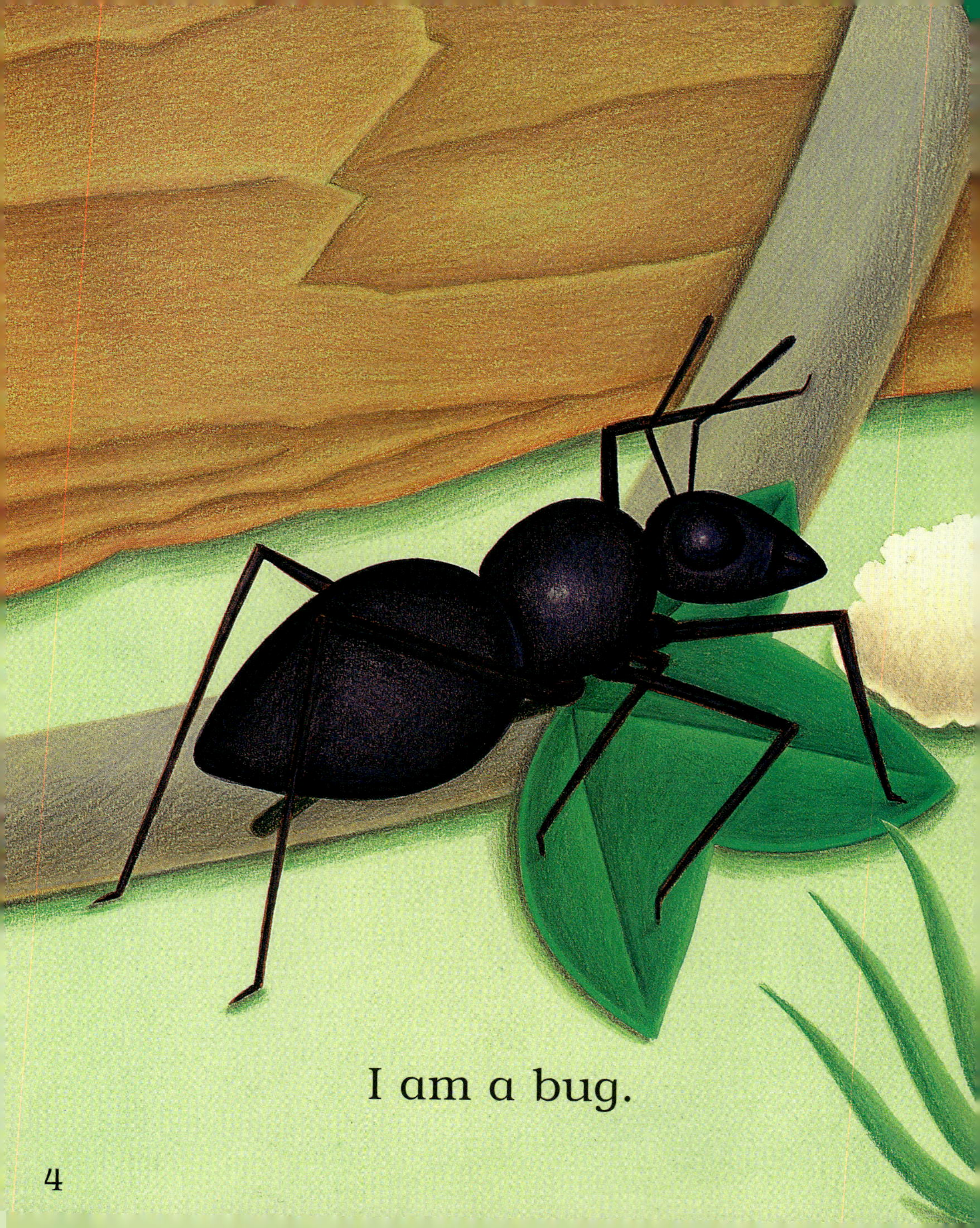

I am a bug.

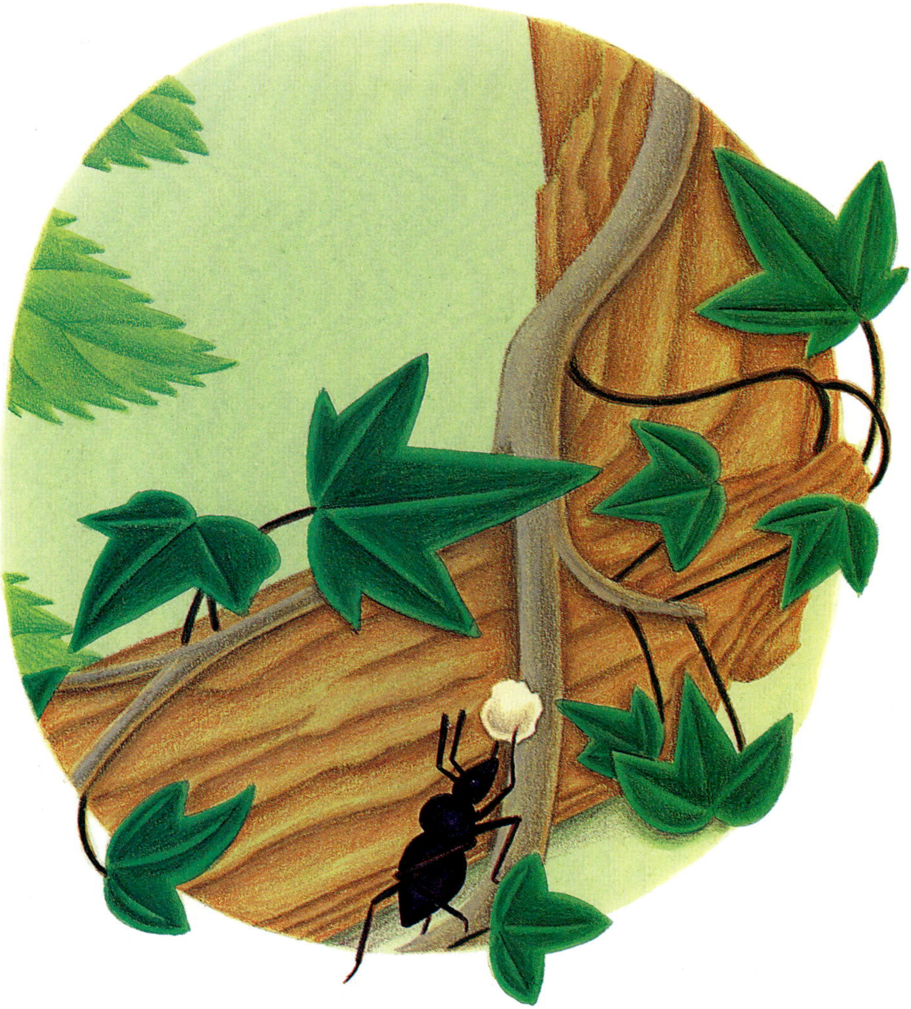

I run up.

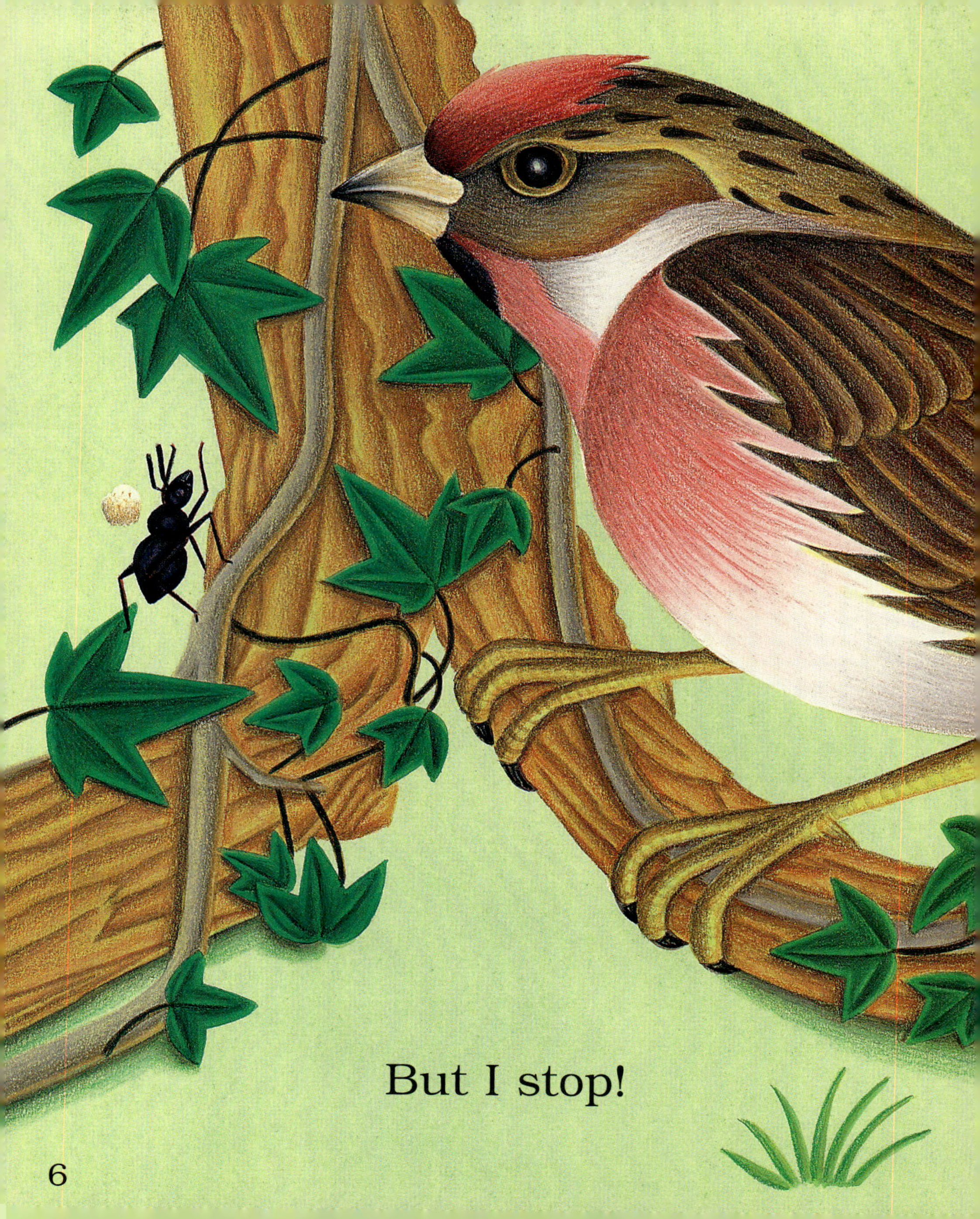

But I stop!

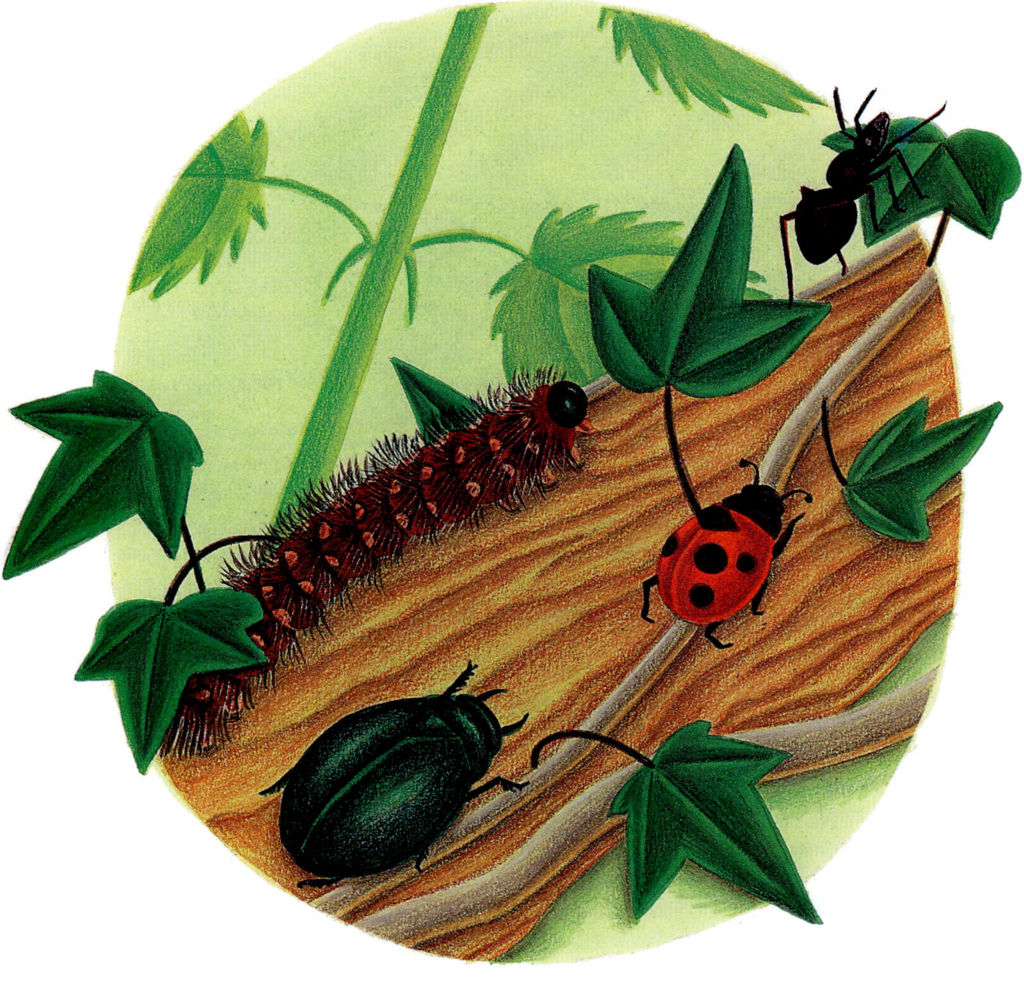

Four bugs run up.

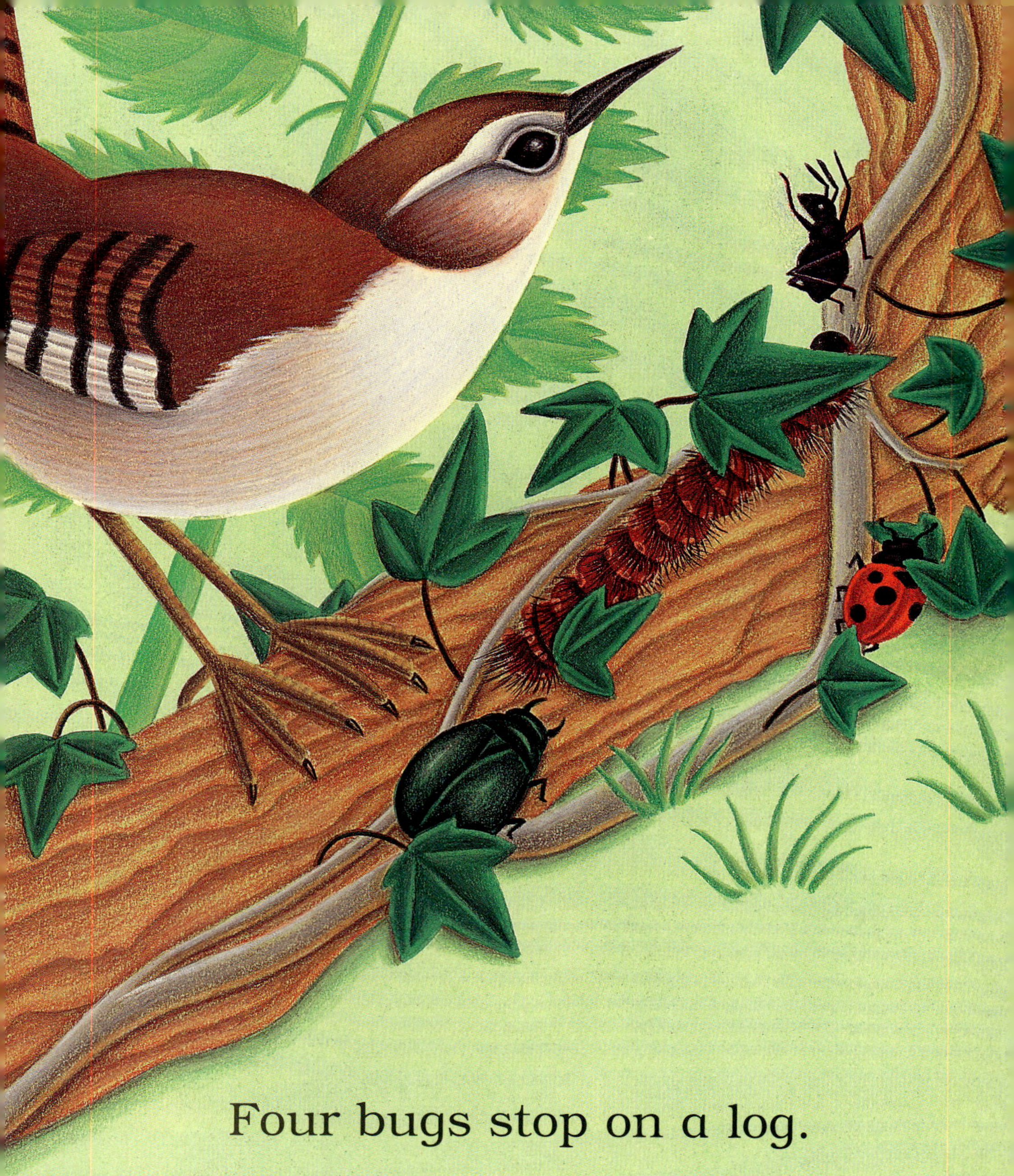

Four bugs stop on a log.